Learn

Eureka Math®
Grade K
Modules 1 & 2

Published by Great Minds®.

Copyright © 2018 Great Minds®.

Printed in the U.S.A.
This book may be purchased from the publisher at eureka-math.org.
CMP   10   9   8   7   6   5   4   3   2

ISBN 978-1-64054-076-7

GK-M1-M2-L-05.2018

Learn ◆ Practice ◆ Succeed

Eureka Math® student materials for *A Story of Units*® (K–5) are available in the *Learn, Practice, Succeed* trio. This series supports differentiation and remediation while keeping student materials organized and accessible. Educators will find that the *Learn, Practice,* and *Succeed* series also offers coherent—and therefore, more effective—resources for Response to Intervention (RTI), extra practice, and summer learning.

Learn

Eureka Math Learn serves as a student's in-class companion where they show their thinking, share what they know, and watch their knowledge build every day. *Learn* assembles the daily classwork—Application Problems, Exit Tickets, Problem Sets, templates—in an easily stored and navigated volume.

Practice

Each *Eureka Math* lesson begins with a series of energetic, joyous fluency activities, including those found in *Eureka Math Practice.* Students who are fluent in their math facts can master more material more deeply. With *Practice,* students build competence in newly acquired skills and reinforce previous learning in preparation for the next lesson.

Together, *Learn* and *Practice* provide all the print materials students will use for their core math instruction.

Succeed

Eureka Math Succeed enables students to work individually toward mastery. These additional problem sets align lesson by lesson with classroom instruction, making them ideal for use as homework or extra practice. Each problem set is accompanied by a Homework Helper, a set of worked examples that illustrate how to solve similar problems.

Teachers and tutors can use *Succeed* books from prior grade levels as curriculum-consistent tools for filling gaps in foundational knowledge. Students will thrive and progress more quickly as familiar models facilitate connections to their current grade-level content.

Students, families, and educators:

Thank you for being part of the *Eureka Math*® community, where we celebrate the joy, wonder, and thrill of mathematics.

In the *Eureka Math* classroom, new learning is activated through rich experiences and dialogue. The *Learn* book puts in each student's hands the prompts and problem sequences they need to express and consolidate their learning in class.

What is in the Learn book?

Application Problems: Problem solving in a real-world context is a daily part of *Eureka Math.* Students build confidence and perseverance as they apply their knowledge in new and varied situations. The curriculum encourages students to use the RDW process—Read the problem, Draw to make sense of the problem, and Write an equation and a solution. Teachers facilitate as students share their work and explain their solution strategies to one another.

Problem Sets: A carefully sequenced Problem Set provides an in-class opportunity for independent work, with multiple entry points for differentiation. Teachers can use the Preparation and Customization process to select "Must Do" problems for each student. Some students will complete more problems than others; what is important is that all students have a 10-minute period to immediately exercise what they've learned, with light support from their teacher.

Students bring the Problem Set with them to the culminating point of each lesson: the Student Debrief. Here, students reflect with their peers and their teacher, articulating and consolidating what they wondered, noticed, and learned that day.

Exit Tickets: Students show their teacher what they know through their work on the daily Exit Ticket. This check for understanding provides the teacher with valuable real-time evidence of the efficacy of that day's instruction, giving critical insight into where to focus next.

Templates: From time to time, the Application Problem, Problem Set, or other classroom activity requires that students have their own copy of a picture, reusable model, or data set. Each of these templates is provided with the first lesson that requires it.

Where can I learn more about Eureka Math *resources?*

The Great Minds® team is committed to supporting students, families, and educators with an ever-growing library of resources, available at eureka-math.org. The website also offers inspiring stories of success in the *Eureka Math* community. Share your insights and accomplishments with fellow users by becoming a *Eureka Math* Champion.

Best wishes for a year filled with aha moments!

Jill Diniz

Jill Diniz
Director of Mathematics
Great Minds

The Read–Draw–Write Process

The *Eureka Math* curriculum supports students as they problem-solve by using a simple, repeatable process introduced by the teacher. The Read–Draw–Write (RDW) process calls for students to

1. Read the problem.

2. Draw and label.

3. Write an equation.

4. Write a word sentence (statement).

Educators are encouraged to scaffold the process by interjecting questions such as

- What do you see?

- Can you draw something?

- What conclusions can you make from your drawing?

The more students participate in reasoning through problems with this systematic, open approach, the more they internalize the thought process and apply it instinctively for years to come.

Contents

Module 1: Numbers to 10

Module 2: Two-Dimensional and Three-Dimensional Shapes

Grade K
Module 1

Draw the sock.

 Draw

(Hold up a sock.) Draw a picture of this sock.

Lesson 1:   Analyze to find two objects that are *exactly the same or not exactly the same*.

© 2018 Great Minds®. eureka-math.org

3

Monday

Name _____   Date _____

Find animals that are exactly the same. Then, find animals that look like each other but are not exactly the same. Use a ruler to draw a line connecting the animals.

EUREKA MATH

Lesson 1:   Analyze to find two objects that are *exactly the same or not exactly the same*.

© 2018 Great Minds®. eureka-math.org

5

Jeremy has 3 marbles.

Draw his marbles.

 Draw

How is your partner's drawing exactly the same? How is your partner's drawing not exactly the same?

Name _____   Date _____

Use your ruler to draw a line between two objects that match.
Use your words. "These are the same, but this one _____, and this
one _____."

Draw two circles that are the same but a different color.

 **Draw**

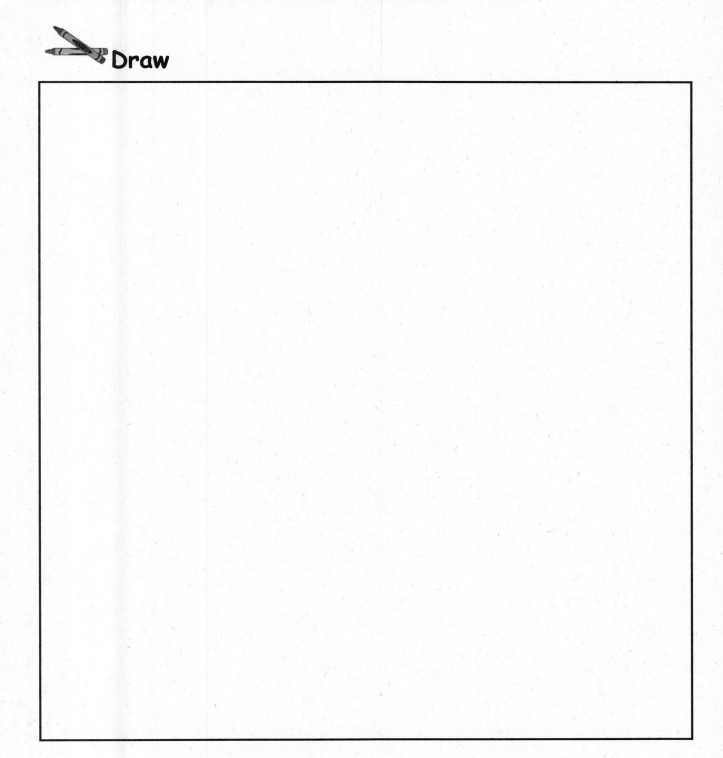

Name _____   Date _____

Draw a line between the objects that have the same pattern. Talk with a neighbor about the objects that match.

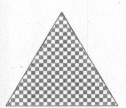

EUREKA MATH®

© 2018 Great Minds®. eureka-math.org

Circle the object that would be used together with the object on the left.

Lesson 3:   Classify to find two objects that share a visual pattern, color, and use.

© 2018 Great Minds®. eureka-math.org

EUREKA
MATH®

Name _____   Date _____

Use the cutouts. Glue the pictures to show where to keep each thing.

Problem Set Cutouts

Sort the class into two groups.

Draw

Draw a picture to show how you sorted the class into two groups.

Lesson 5:   Classify items into three categories, determine the count in each, and reason
about how the last number named determines the total.

© 2018 Great Minds®. eureka-math.org

21

Name _____   Date _____

Draw a line with your ruler to show where each thing belongs.

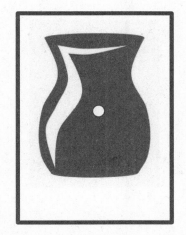
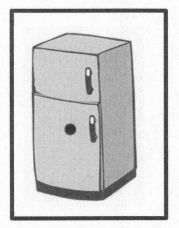
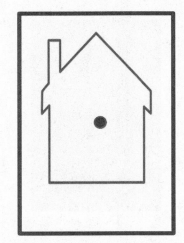

EUREKA MATH

Lesson 5:   Classify items into three categories, determine the count in each, and reason about how the last number named determines the total.

23

Draw 1 thing you would wear in the summer.

Draw 1 thing you would wear in the winter.

Draw

Tell a friend about your drawing.

Lesson 6:   Sort categories by count. Identify categories with 2, 3, and 4 within a given scenario.

© 2018 Great Minds®. eureka-math.org

25

Name _____   Date _____

Count and color.

2 red	3 blue	4 orange

EUREKA MATH

Lesson 6:   Sort categories by count. Identify categories with 2, 3, and 4 within a given scenario.

© 2018 Great Minds®. eureka-math.org

27

Find two things in this room.

Draw what you found.

Draw

Show your friend. How many things did you and your friend find all together? Did you find some of the same things? If so, put them together and count them.

Lesson 7:   Sort by count in vertical columns and horizontal rows (linear configurations to 5). Match to numerals on cards.

© 2018 Great Minds®. eureka-math.org

29

EUREKA MATH®

Name _____ Date _____

Count and color.

| 1 | 2 | 3 | 4 | 5 |
| black | blue | brown | red | yellow |

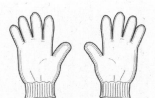

EUREKA MATH

Lesson 7: Sort by count in vertical columns and horizontal rows (linear configurations to 5). Match to numerals on cards.

31

© 2018 Great Minds®. eureka-math.org

Put 4 counters in a row going across.

Put 4 counters in a column going up and down.

 Draw

(Give each student counters in a bag.) Draw your counters on your paper.

Name _____ Date _____

Count the objects. Circle the correct number.

1   2   3

1   2   3

3   4   5

2   3   4

4   3   2

5   4   1

4   3   2

5   4   1

Lesson 8: Answer *how many* questions to 5 in linear configurations (5-group), with 4 in an array configuration. Compare ways to count five fingers.

35

EUREKA MATH

Draw a caterpillar pet that has 4 different parts.

Draw

Show your pet to a friend.

Lesson 9:   Within linear and array dot configurations of numbers 3, 4, and 5, find
              hidden partners.

© 2018 Great Minds®. eureka-math.org

37

Name _____ Date _____

Count the dots, and circle the correct number. Color the same number of dots on the right as the gray ones on the left to show the hidden partners.

3   4   5

3   4   5

3   4   5

3   4   5

Lesson 9:   Within linear and array dot configurations of numbers 3, 4, and 5, find
              hidden partners.

© 2018 Great Minds®. eureka-math.org

39

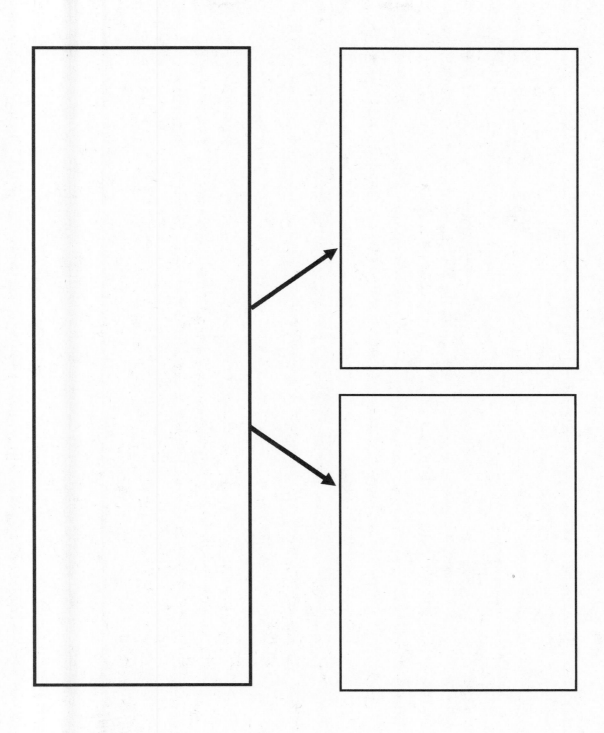

hidden partners

Lesson 9:   Within linear and array dot configurations of numbers 3, 4, and 5, find
hidden partners.

41

Name _____   Date _____

Color to see the hidden partners.

Count the objects. Circle the total number.

Color 1 circle.	Color 1 star.	Color 1 circle.
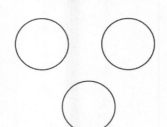		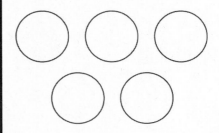
1 2 3	2 3 4	3 4 5
Color 2 stars.	Color 2 circles.	Color 2 stars.
	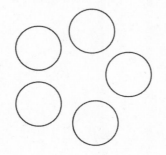	
3 2 1	5 4 3	4 5 3

Draw 2 circles. Count all the objects, and circle the number.

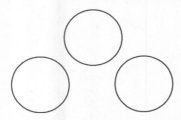

                                        5   2   3

EUREKA MATH   **Lesson 10:**   Within circular and scattered dot configurations of numbers 3, 4, and 5, find *hidden partners*   45

© 2018 Great Minds®. eureka-math.org

There are 4 crayons on the floor.

Draw the crayons.

 Draw

How many crayons are the same color as your friend's?

Lesson 11:   Model decompositions of 3 with materials, drawings, and expressions.
Represent the decomposition as 1 + 2 and 2 + 1.

© 2018 Great Minds®. eureka-math.org

47

Name _____   Date _____

Count the squares. Draw the squares above the numbers.

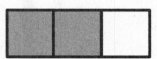

2   +   1

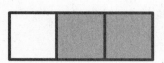

1   +   2

3   +   1

1   +   3

4   +   1

1   +   4

EUREKA MATH

Lesson 11:   Model decompositions of 3 with materials, drawings, and expressions. Represent the decomposition as 1 + 2 and 2 + 1.

© 2018 Great Minds®. eureka-math.org

49

Draw 4 apples.

Color some red and some green.

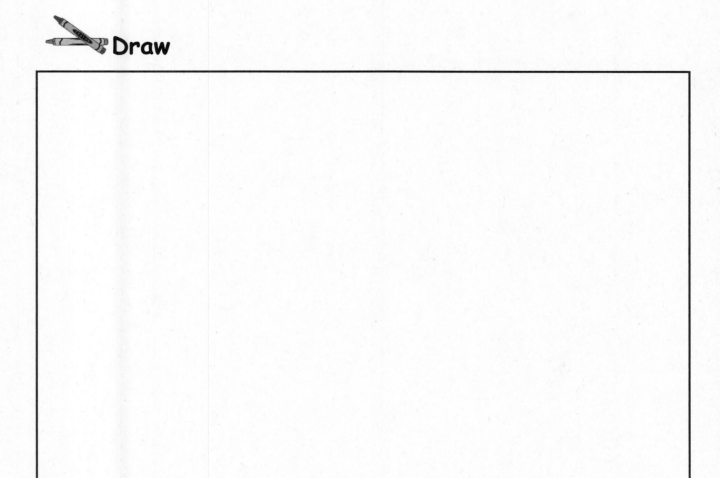 **Draw**

Tell your friend how many apples are red and how many are green. Did you and your friend have the same number of red apples?

Lesson 12:   Understand the meaning of zero. Write the numeral 0.   **51**

© 2018 Great Minds®. eureka-math.org

Name _____   Date _____

Write 0.

numeral formation practice sheet 0

Lesson 12:   Understand the meaning of zero. Write the numeral 0.   **53**

© 2018 Great Minds®. eureka-math.org

Name _____   Date _____

Circle the number that tells how many.

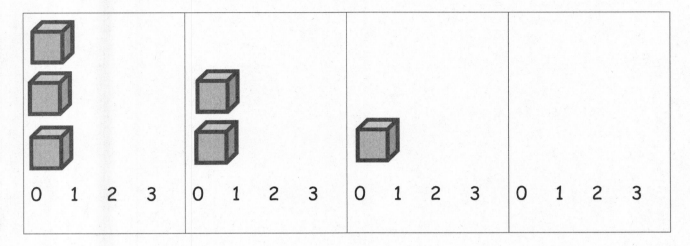

- -

How many elephants are in the trees?

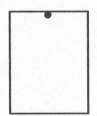

EUREKA
MATH

Johnny had 2 cookies.

He gave 1 to a friend and ate 1 himself.
How many cookies does he have now?

 **Draw**

Lesson 13:   Order and write numerals 0–3 to answer *how many* questions.

57

© 2018 Great Minds®. eureka-math.org

Name _____ Date _____

Write 1, 2, and 3.

EUREKA MATH

Lesson 13: Order and write numerals 0–3 to answer *how many* questions.

59

© 2018 Great Minds®. eureka-math.org

Name _____ Date _____

Write the missing numbers.

1 2 ☐	3 2 ☐
1 ☐ 3	3 ☐ 1
0 1 ☐	☐ 2 1
0 ☐ 2	2 1 ☐
☐ 1 2	☐ 2 0

Count and write how many.

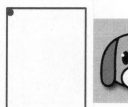

Lesson 13: Order and write numerals 0–3 to answer *how many* questions.

© 2018 Great Minds®. eureka-math.org

EUREKA MATH

How many?

 Write

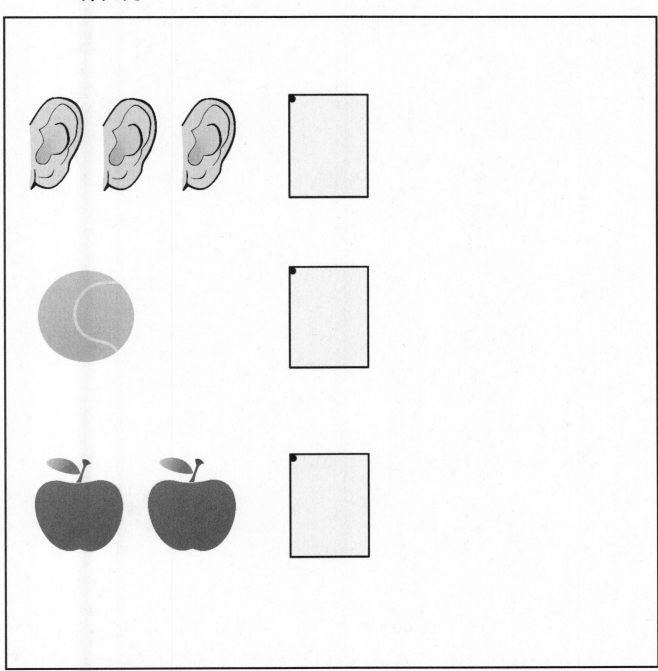

Write the number in the box.

Lesson 14:   Write numerals 1–3. Represent decompositions with materials, drawings, and equations, 3 = 2 + 1 and 3 = 1 + 2.

© 2018 Great Minds®. eureka-math.org

63

Name _____   Date _____

Color the picture to match the number sentence.

$$3 = 1 + 2$$

$$3 = 2 + 1$$

$$3 = 1 + 2$$

$$3 = 2 + 1$$

Write how many.

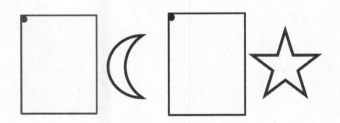

Write how many.

EUREKA MATH

Lesson 14:   Write numerals 1–3. Represent decompositions with materials,
drawings, and equations, 3 = 2 + 1 and 3 = 1 + 2.

65

Draw 3 circles.

Color 2 blue and 1 red.

 **Draw**

Write

Complete the number sentence:

$$3 = \boxed{} + \boxed{}$$

Lesson 15: Order and write numerals 4 and 5 to answer *how many* questions in categories; sort by count.

67

Name _____   Date _____

Write 4 and 5.

Write the missing numbers:

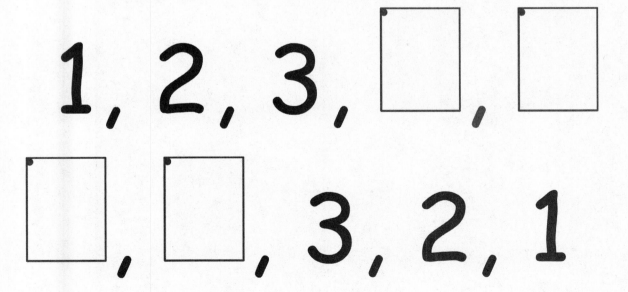

Lesson 15:   Order and write numerals 4 and 5 to answer *how many* questions in categories; sort by count.

© 2018 Great Minds®. eureka-math.org

69

Name _____   Date _____

Count and write how many. Circle a group of four of each fruit.

Lesson 15:   Order and write numerals 4 and 5 to answer *how many* questions in
            categories; sort by count.

© 2018 Great Minds®. eureka-math.org

71

Draw 4 cups and 5 straws.

Draw

Write how many cups. Write how many straws. Circle the number that is more.

Lesson 16: Write numerals 1–5 in order. Answer and make drawings of decompositions with totals of 4 and 5 without equations.

© 2018 Great Minds®. eureka-math.org

73

Name _____   Date _____

In each picture, color some squares red and some blue. Do it a different way each time.

How many squares?

How many squares?

How many squares?

How many squares?

Draw more circles to make 4.

◯◯◯   ◯◯   ◯

Draw more X's to make 5.

XXXX   XXX   XX   X

Lesson 16:   Write numerals 1–5 in order. Answer and make drawings of decompositions with totals of 4 and 5 without equations.

© 2018 Great Minds®. eureka-math.org

75

Name _____   Date _____

How many 🌙? ☐

How many 🌙? ☐

How many altogether? ☐

How many 🙂? ☐

How many 🙁? ☐

How many altogether? ☐

EUREKA MATH

Lesson 16: Write numerals 1–5 in order. Answer and make drawings of decompositions with totals of 4 and 5 without equations.

77

© 2018 Great Minds®. eureka-math.org

Finish this sentence:

I could eat 5 _____.

 Draw

Draw a picture to show your idea.

EUREKA MATH   **Lesson 17:**   Count 4–6 objects in vertical and horizontal linear configurations and   **79**

array configurations. Match 6 objects to the numeral 6.

© 2018 Great Minds®. eureka-math.org

Name _____   Date _____

Draw 1 more. Then count the objects, and write the number in the box.

How many?   ☐

How many?   ☐

How many?   ☐

Draw 1 more.
Then, circle the number.

Draw 6 fingers.

Draw 6 beads.

4   5   6

Lesson 17:   Count 4–6 objects in vertical and horizontal linear configurations and array configurations. Match 6 objects to the numeral 6.

© 2018 Great Minds®. eureka-math.org

81

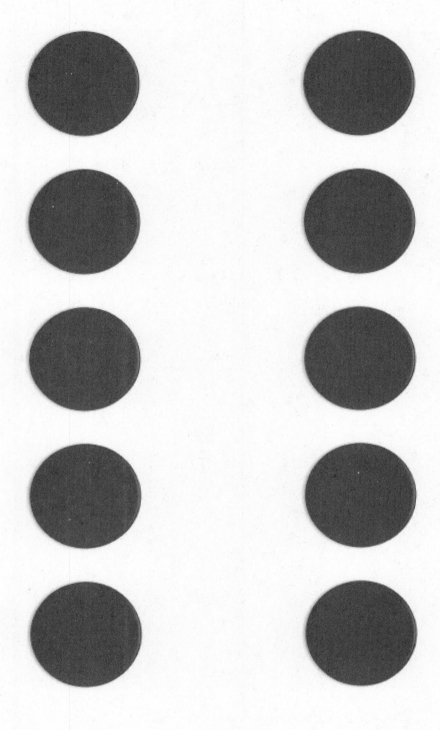

5-group mat

Lesson 17: Count 4–6 objects in vertical and horizontal linear configurations and array configurations. Match 6 objects to the numeral 6.

83

© 2018 Great Minds®. eureka-math.org

Make a row of 3 dots.

Make another row of 3 dots under the first one.

Draw

Count your dots. Tell your friend how many.

Lesson 18:   Count 4–6 objects in circular and scattered configurations. Count 6 items out of a larger set. Write numerals 1–6 in order.

85

© 2018 Great Minds®. eureka-math.org

Name _____   Date _____

Write 6.

6 ☐ ☐ ___ ___

6 ☐ ☐ ___ ___

Write the missing numbers:

1, 2, 3, 4, ☐, ☐,

☐, ☐, 4, 3, 2, 1

_____
numeral formation practice sheet 6

Lesson 18: Count 4–6 objects in circular and scattered configurations. Count 6 items out of a larger set. Write numerals 1–6 in order.

© 2018 Great Minds®. eureka-math.org

87

Draw 5 ice cream cones.

Draw 1 more ice cream cone.

Draw

Count how many ice cream cones you drew on your paper. Write the number.

Lesson 19:   Count 5–7 linking cubes in linear configurations. Match with numeral 7. Count on fingers from 1 to 7, and connect to 5 group images.

© 2018 Great Minds®. eureka-math.org

91

Name _____   Date _____

Color 5 in each group.

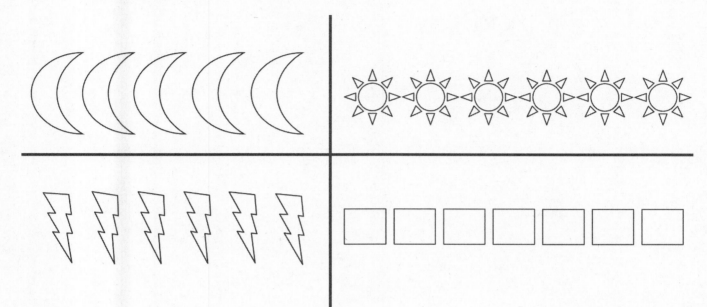

Color 5. Draw 2 circles to the right. Count all the circles.

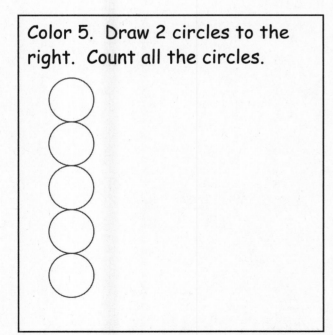

Color 5. Draw 2 circles below. Count all the circles.

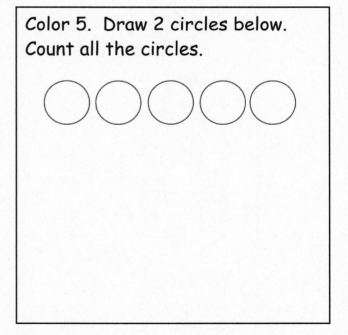

EUREKA MATH®

Lesson 19:   Count 5–7 linking cubes in linear configurations. Match with numeral 7. Count on fingers from 1 to 7, and connect to 5 group images.

© 2018 Great Minds®. eureka-math.org

93

Christopher has a bag of 5 cookies and 2 other loose cookies.

Draw

Draw the cookies. Count the cookies with your partner. How many cookies does Christopher have?

Name _____   Date _____

Write 7.

Write the missing numbers:

☐ , 2, 3, 4, 5, ☐ , ☐

7, 6, ☐ , 4, 3, ☐ , ☐

Lesson 20: Reason about sets of 7 varied objects in circular and scattered configurations. Find a path through the scattered configuration. Write numeral 7. Ask, "How is your seven different from mine?"

© 2018 Great Minds®. eureka-math.org

97

Name _____   Date _____

Color 7 beans. Draw a line to connect the beans you colored.

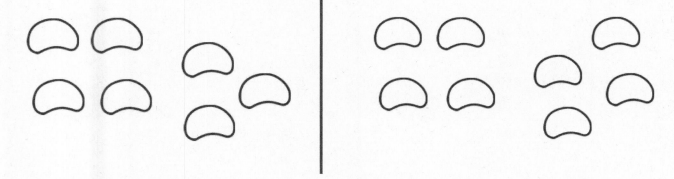

Color 7 beans.

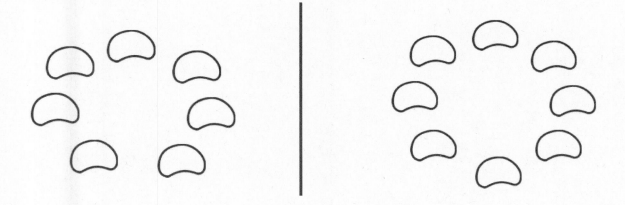

Count the dots in each box. Write the number.

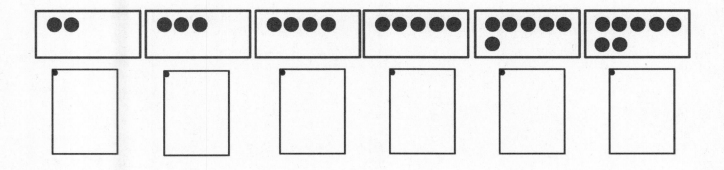

EUREKA
MATH®

Lesson 20:   Reason about sets of 7 varied objects in circular and scattered configurations.
Find a path through the scattered configuration. Write numeral 7. Ask, "How
is your seven different from mine?"

© 2018 Great Minds®. eureka-math.org

99

Some children were playing with marbles.

Draw 7 marbles.

Draw

Count the marbles with your friend. What would happen if someone gave the children another marble?

Lesson 21:   Compare counts of 8 in linear and array configurations. Match with numeral 8.

101

Draw 2 stacks of 4 blocks each.

Draw

Count your blocks. How many do you have? How many does your friend have?

Lesson 22:   Arrange and strategize to count 8 beans in circular (around a cup) and scattered configurations. Write numeral 8. Find a path through the scatter set, and compare paths with a partner.

© 2018 Great Minds®. eureka-math.org

EUREKA MATH®   105

Name _____   Date _____

Write 8.

Color 8 happy faces.

Circle a different group of 8 happy faces.

Lesson 22: Arrange and strategize to count 8 beans in circular (around a cup) and scattered configurations. Write numeral 8. Find a path through the scatter set, and compare paths with a partner.

© 2018 Great Minds®. eureka-math.org

Name _____   Date _____

Draw a counting path with a line to show the order in which you counted.
Write the total number. Circle a group of 5 in each set.

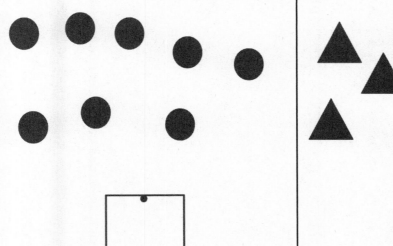

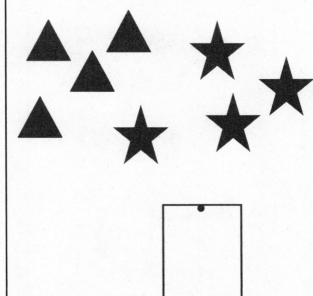

Color 8 circles.

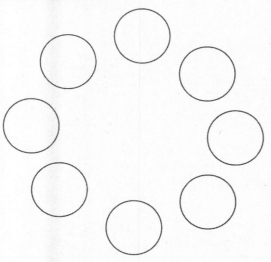

Color 8 circles.

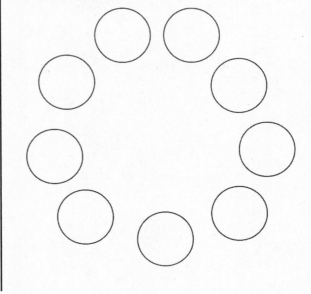

EUREKA MATH®

Lesson 22:   Arrange and strategize to count 8 beans in circular (around a cup) and scattered
configurations. Write numeral 8. Find a path through the scatter set, and
compare paths with a partner.

© 2018 Great Minds®. eureka-math.org

109

Draw a circle with 8 balls inside.

Draw

Count the balls. Share your counting with a friend.

Lesson 23:   Organize and count 9 varied geometric objects in linear and array (3 threes) configurations. Place objects on 5-group mat. Match with numeral 9.

Name _____   Date _____

Count and circle how many. Color 5.

7   8   9

7   8   9

Draw 4 circles to the right. Count all the circles.	Draw 4 circles below. Count all the circles.

EUREKA MATH

Lesson 23:   Organize and count 9 varied geometric objects in linear and array (3 threes) configurations. Place objects on 5-group mat. Match with numeral 9.

113

© 2018 Great Minds®. eureka-math.org

Color 3. Count and circle how many.

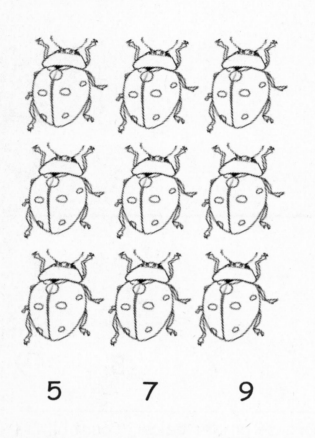

5   7   9   6   8   9

Color 3. Draw 2 circles to finish the last row. Count how many.

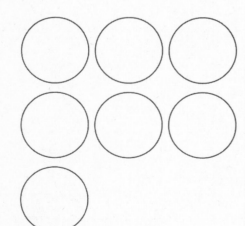

Lesson 23:   Strategize to count 9 objects in circular (around a paper plate) and scattered configurations printed on paper. Write numeral 9. Represent a path through the scatter count with each object.

© 2018 Great Minds®. eureka-math.org

EUREKA MATH

Draw 5 shapes.

Draw 4 more shapes.

How many shapes are there?

Draw

Lesson 24:   Strategize to count 9 objects in circular (around a paper plate) and scattered
configurations printed on paper. Write numeral 9. Represent a path through
the scatter count with each object.

© 2018 Great Minds®. eureka-math.org

115

Name _____   Date _____

Write 9.

Color 9 happy faces.

Circle a different group of 9 happy faces.

_____

numeral formation practice sheet 9

Lesson 24:   Strategize to count 9 objects in circular (around a paper plate) and scattered
configurations printed on paper. Write numeral 9. Represent a path through
the scatter count with each object.

© 2018 Great Minds®. eureka-math.org

117

Name _____   Date _____

Draw a counting path with a line to show the order in which you counted.
Write the total number. Circle a group of 5 in each set.

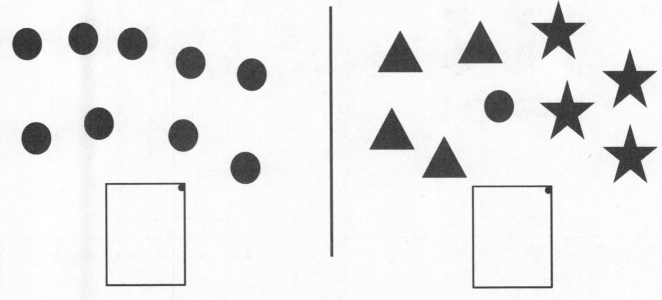

Count the stars and objects. Write the total number of objects
in the boxes.

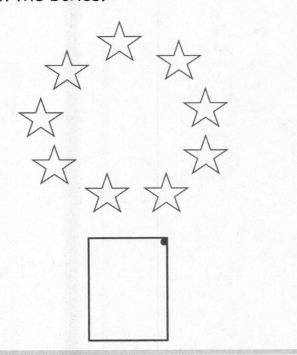

Lesson 24:   Count 10 objects in linear and array configurations (2 fives). Match with
numeral 10. Place on the 5-group mat. Dialogue about 9 and 10.
Write numeral 10.

© 2018 Great Minds®. eureka-math.org

119

Count the dots.
Write the number.

Count the dots. Write the number.
Circle a group of 5.

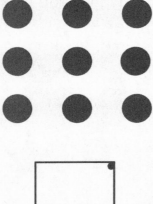

Color 5. Count all the circles.

Color 5. Draw more circles to make 9.

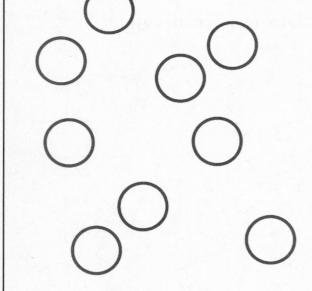

Lesson 24:   Count 10 objects in linear and array configurations (2 fives). Match with
numeral 10. Place on the 5-group mat. Dialogue about 9 and 10.
Write numeral 10.

EUREKA
MATH®

Draw 9 smiley faces.

Write the number 9.

 **Draw**

Count the smiley faces by connecting them with lines. Show your picture to a friend. What would happen if you drew another smiley face in your picture?

 EUREKA MATH®

Lesson 25:   Count 10 objects in linear and array configurations (2 fives). Match with
                numeral 10. Place on the 5-group mat. Dialogue about 9 and 10.
                Write numeral 10.
© 2018 Great Minds®. eureka-math.org

121

Name _____   Date _____

Count and circle how many. Color 5.

8   9   10

8   9   10

| Color 5 circles. Draw 5 circles to the right. Count all the circles. | Color 5 circles. Draw 5 circles below. Count all the circles. |

Lesson 25:   Count 10 objects in linear and array configurations (2 fives). Match with numeral 10. Place on the 5-group mat. Dialogue about 9 and 10. Write numeral 10.

123

© 2018 Great Minds®. eureka-math.org

Count and circle how many. Color 5.

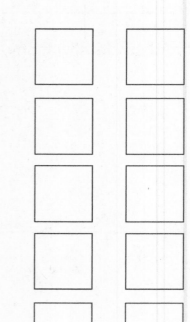

6   8   10   7   9   10

Color 5 circles. Draw 4 circles to finish the row.

Lesson 25:   Count 10 objects in linear and array configurations (2 fives). Match
with numeral 10. Place on the 5-group mat. Dialogue about
9 and 10. Write numeral 10.
© 2018 Great Minds®. eureka-math.org

Draw a row of 5 bricks.

Draw another row of 5 bricks on top.

How many bricks did you draw?

 **Draw**

Lesson 26:   Count 10 objects in linear and array configurations (2 fives). Match
with numeral 10. Place on the 5-group mat. Dialogue about 9 and 10.
Write numeral 10.

125

 EUREKA MATH®

Name _____   Date _____

Write 10.

numeral formation practice sheet 10

Lesson 26:   Count 10 objects in linear and array configurations (2 fives). Match
with numeral 10. Place on the 5-group mat. Dialogue about 9 and 10.
Write numeral 10.

© 2018 Great Minds®. eureka-math.org

127

Name _____   Date _____

Draw 10 circles in a row. Color the first 5 yellow and the second 5 blue.
Write how many circles in the boxes.

Draw 5 circles in the gray part. Draw 5 circles in the white part.
Write how many circles in the boxes.

Draw two towers of 5 next to each other. Color 1 tower red and the other tower orange.	Draw a row of 5 cubes. Draw another row of 5. Count.

Draw a picture of your bracelet on the back of your paper.

Lesson 26:   Count 10 objects in linear and array configurations (2 fives). Match
with numeral 10. Place on the 5-group mat. Dialogue about 9 and 10.
Write numeral 10.

© 2018 Great Minds®. eureka-math.org

129

Make a snowman 5 snowballs high.

Make a snowman next to him that is also 5 snowballs high.

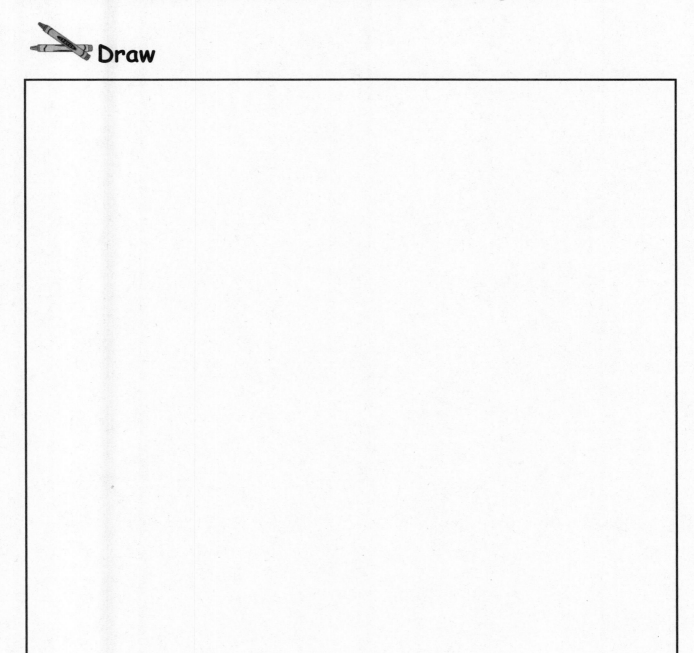 Draw

How many snowballs did you draw? Write the number.

Name _____   Date _____

Count the shapes, and write how many. Color the shape you counted first.

Draw 10 things. Color 5 of them. Color 5 things a different color.

Draw 10 circles. Color 5 circles. Color 5 circles a different color.

Color 10 apples. Draw a path to connect the apples starting at 1.

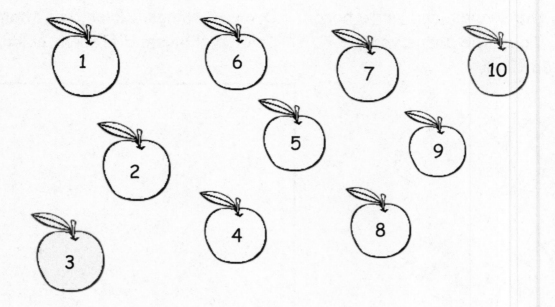

Color 10 apples. Count and draw a path to connect the apples.

EUREKA MATH

Draw a bracelet with 10 beads.

Draw

Have a friend count your beads. Did you both count them the same way? Are there any smaller numbers inside your bracelet?

Name _____   Date _____

Listen to my stories. Color the pictures to show what is happening. Write how many in the box.

Bobby picked 4 red flowers. Then, he picked 2 purple flowers. How many flowers did Bobby pick?

Janet went to the donut store. She bought 6 chocolate donuts and 3 strawberry donuts. How many donuts did she buy?

Some children were sitting in a circle. 4 of them were wearing green shirts. The rest were wearing yellow shirts. How many children were in the circle?

Jerry spilled his bag of marbles. Circle the group of grey marbles. Circle the group of black marbles. How many marbles were spilled?

Make up a story about the bears. Color the bears to match the story.
Tell your story to a friend.

Make up a new story. Draw a picture to go with your story. Tell your
story to a friend.

Lesson 28:   Act out *result unknown* story problems without equations.

EUREKA MATH

Draw 10 little dishes on your paper.

Write the numbers 1–10 on your dishes.

 **Draw**

Count your friend's dishes. Did you both number them the same?

Lesson 29: Order and match numeral and dot cards from 1 to 10. State 1 more than a given number.

139

© 2018 Great Minds®. eureka-math.org

Name _____   Date _____

Count the dots. Write how many. Draw the same number of dots below, but going up and down instead of across. The number 4 has been done for you.

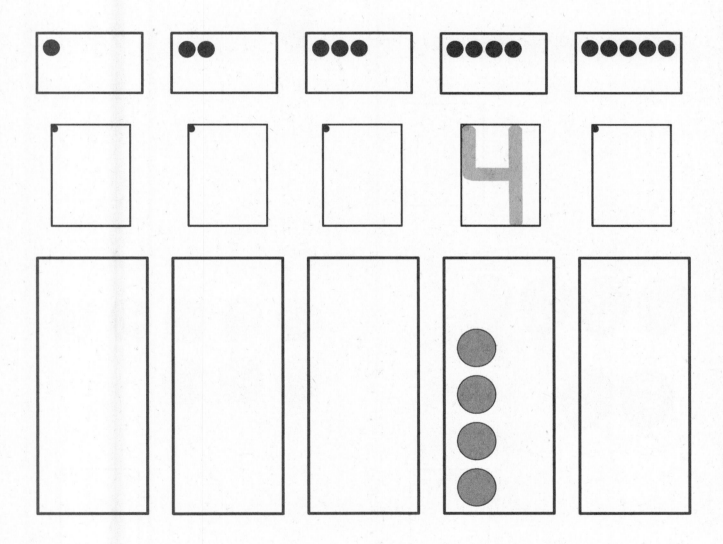

EUREKA MATH

Lesson 29:   Order and match numeral and dot cards from 1 to 10. State 1 more
             than a given number.

© 2018 Great Minds®. eureka-math.org

141

Count the objects. Draw 1 more object. Count and write how many.

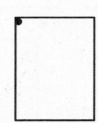

Lesson 29:   Order and match numeral and dot cards from 1 to 10. State 1 more
than a given number.

EUREKA
MATH®

Draw 4 flowers in a vase.

Draw 1 more flower in your vase.

 **Draw**

Count the flowers in your vase. Write the number.

Lesson 30:   Make math stairs from 1 to 10 in cooperative groups.

143

© 2018 Great Minds®. eureka-math.org

Name _____ Date _____

Count and color the white squares red. Count all the cubes in each step.
Write the missing numbers below each step.

1 2 3 ☐ 5 6 ☐ 8 9 ☐ ☐

☐ 2 3 4 5 ☐ 7 8 ☐ 10

Lesson 30:   Make math stairs from 1 to 10 in cooperative groups.

145

EUREKA MATH

Draw a plate of 7 oranges.

Draw 1 more orange.

Draw

How many oranges are on the plate? Write the number.   Tell your friend: There were 7 oranges.

One more is (___).

Lesson 31:   Arrange, analyze, and draw 1 more up to 10 in configurations other
              than towers.

© 2018 Great Minds®. eureka-math.org

147

Name _____   Date _____

Color and count the empty circles. Count the gray circles. Write how many gray circles in the box.

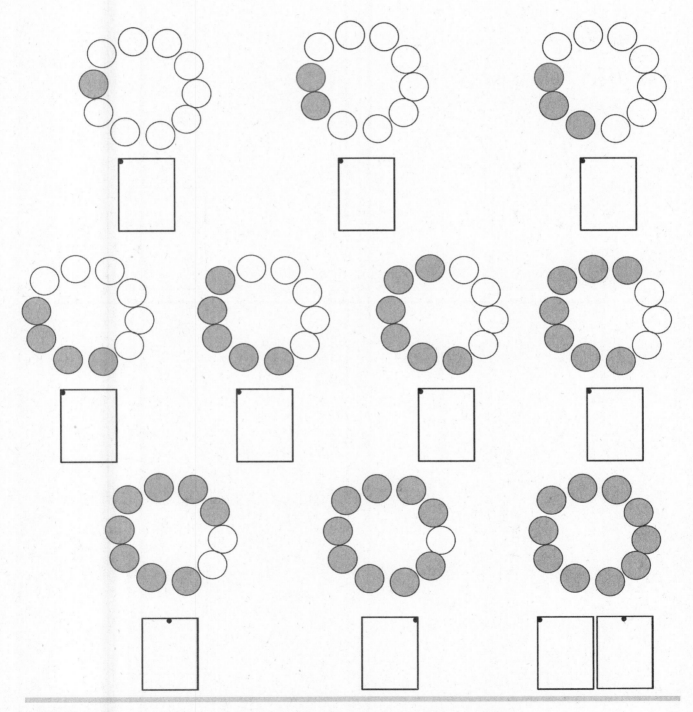

Lesson 31:   Arrange, analyze, and draw 1 more up to 10 in configurations other than towers.

© 2018 Great Minds®. eureka-math.org

149

Draw 1 more circle and count all the circles. Write how many.

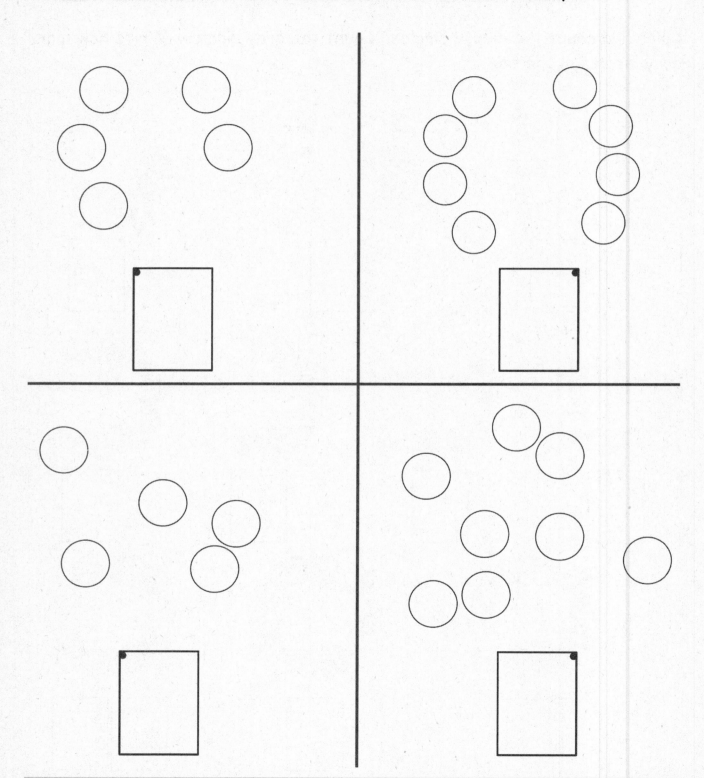

Lesson 31:   Arrange, analyze, and draw 1 more up to 10 in configurations other than towers.

© 2018 Great Minds®. eureka-math.org

The numbers on some of the shirts washed off.

Write the numbers on the shirts!

 Draw

Name _____   Date _____

Draw and write the number of the missing steps.

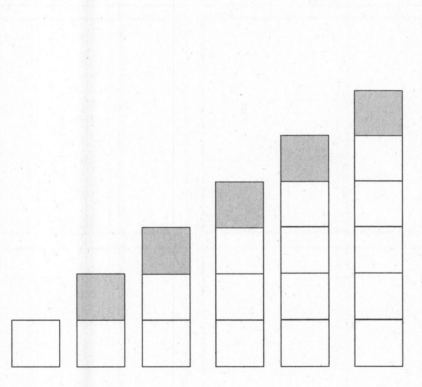

1   2   3   4   5   6   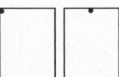   10

Lesson 32:   Arrange, analyze, and draw sequences of quantities of 1 more, beginning with numbers other than 1.

153

Write the missing number. Draw objects to show the numbers.

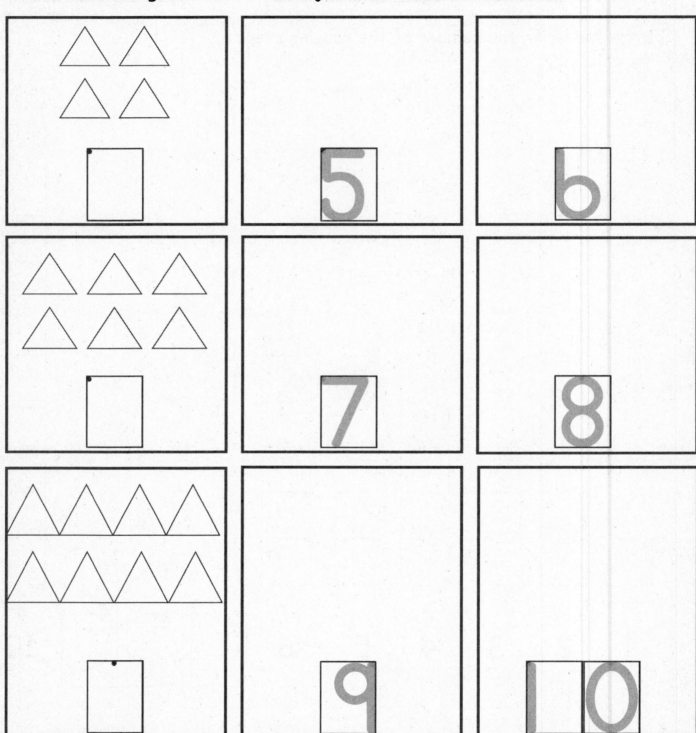

Lesson 32: Arrange, analyze, and draw sequences of quantities of 1 more, beginning with numbers other than 1.

© 2018 Great Minds®. eureka-math.org

EUREKA MATH

Margaret baked some biscuits.

Her kitten jumped on the table and took one away.

Draw an X on one biscuit to show the biscuit the kitten took.

 **Draw**

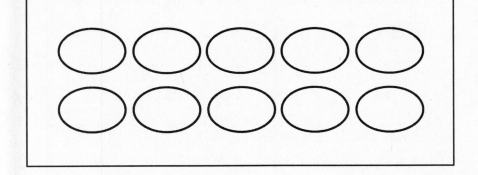

How many are there now? Write the number.

Name _____ Date _____

Count the dots. Write how many. Draw the same number of dots below, but go up. The number 6 is done for you.

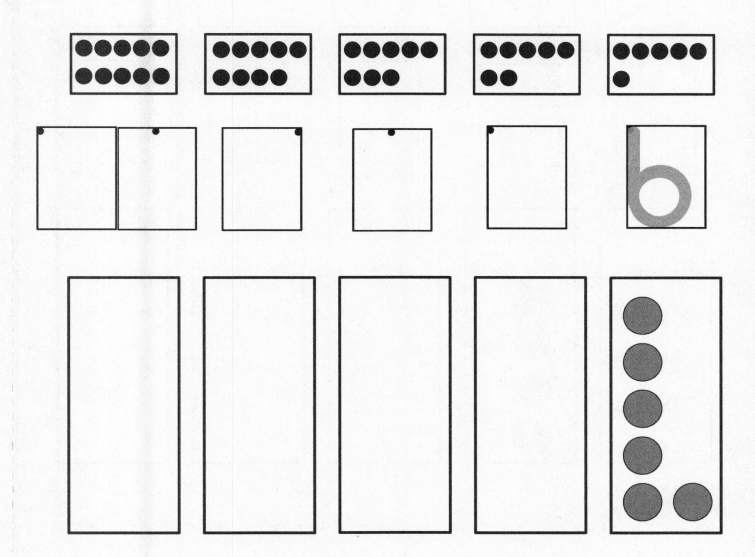

Lesson 33:   Order quantities from 10 to 1, and match numerals.

157

Count the dots. Write how many. Draw the same number of dots below, but go up. The number 4 is done for you.

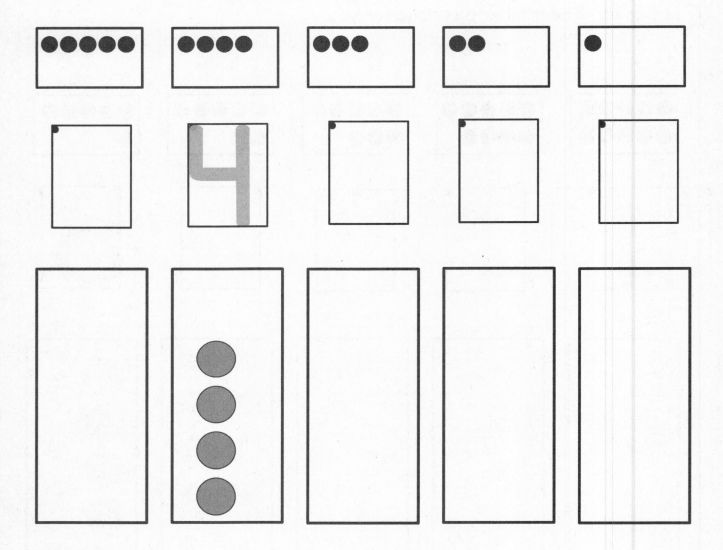

Lesson 33:   Order quantities from 10 to 1, and match numerals.

EUREKA
MATH

Count the balloons. Cross out 1 balloon. Count and write how many balloons are left in the box.

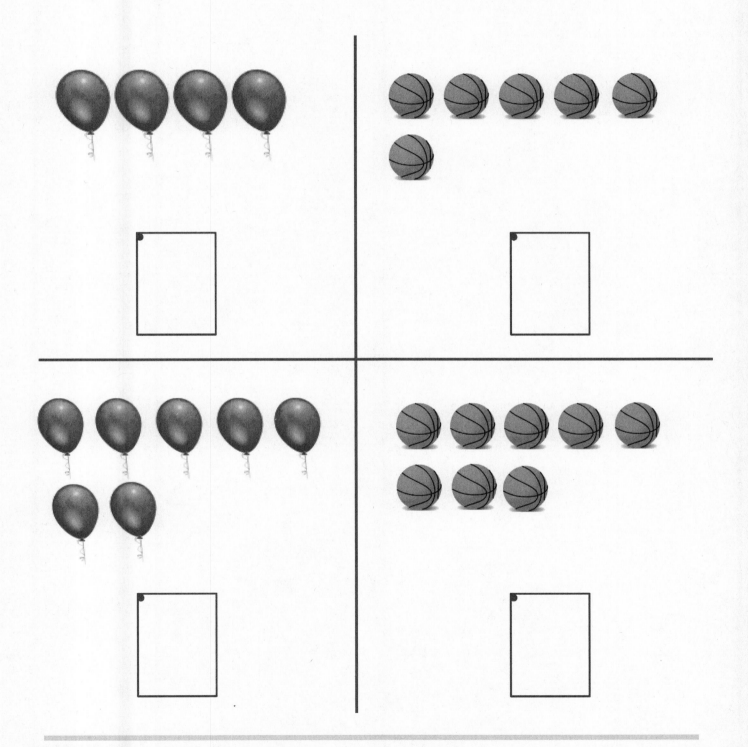

Lesson 33:   Order quantities from 10 to 1, and match numerals.

159

EUREKA MATH

Draw 8 grapes on the first plate. Draw 1 less on the next plate.

Draw 6 straws in the first cup. Draw 1 less in the next cup.

 Draw

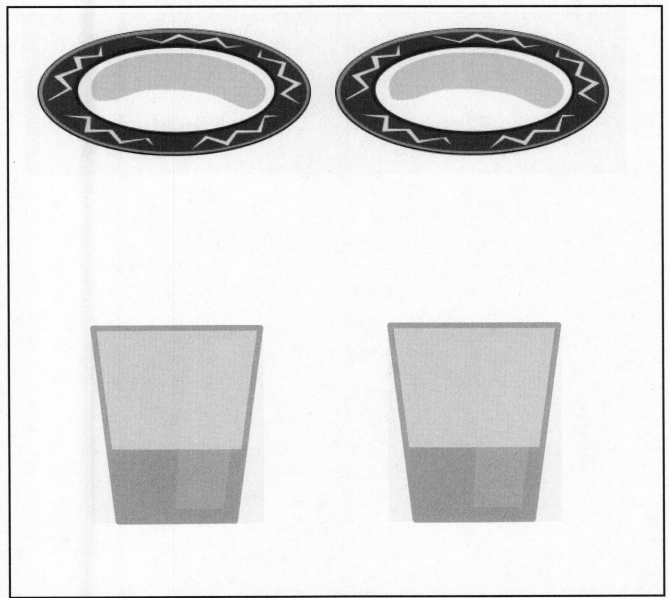

Write how many grapes there are below each plate. Write how many straws there are below each cup.

Lesson 34:   Count down from 10 to 1, and state 1 less than a given number.

161

© 2018 Great Minds®. eureka-math.org

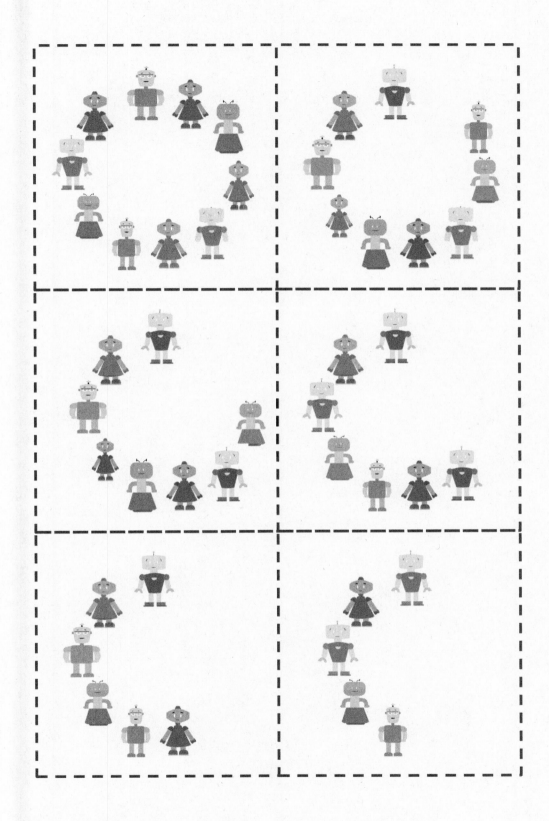

EUREKA MATH

Lesson 34:   Count down from 10 to 1, and state 1 less than a given number.

163

© 2018 Great Minds®. eureka-math.org

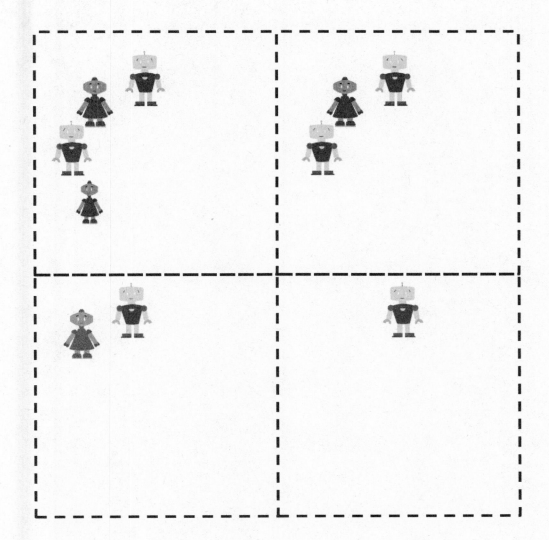

1	2	3	4	5
6	7	8	9	10

Draw a snow girl that is 3 snowballs high.
Next to her, draw a snow boy with 1 less.
How many snowballs are in your snow boy?

 Draw

Compare your picture with a friend's picture.

Name _____   Date _____

Count all the squares in each tower, and write how many. What do you notice?

10

EUREKA MATH

© 2018 Great Minds®. eureka-math.org

Count the number of squares in each stair. Cross off the top square. Use your words to say, "10. One less is nine. 9. One less is eight." Write how many squares are in each stair after you cross off.

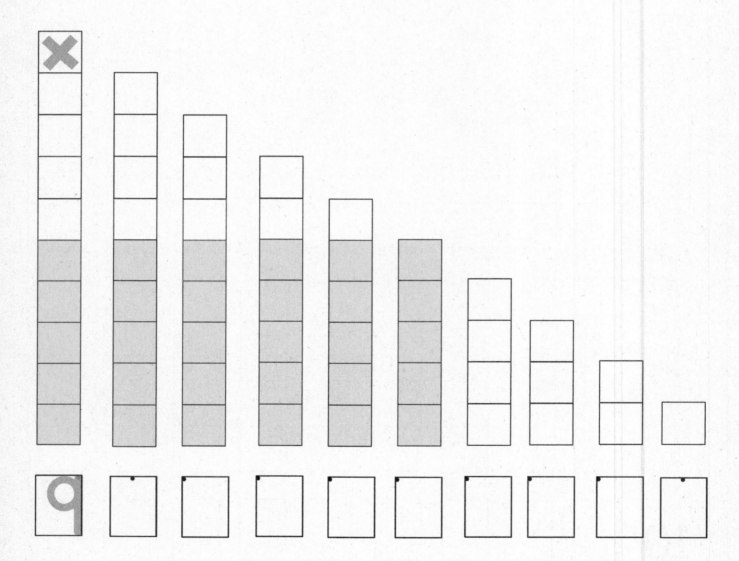

Lesson 35:   Arrange number towers in order from 10 to 1, and describe the pattern.

EUREKA MATH

These towers are mixed up!
Draw the towers in order so each tower shows 1 less.

 **Draw**

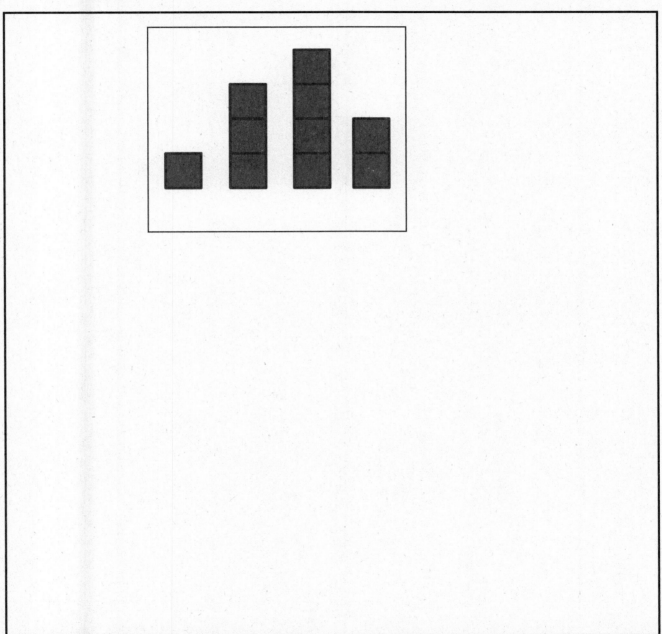

Write the numbers underneath the towers.

Lesson 36:   Arrange, analyze, and draw sequences of quantities that are 1 less
configurations other than towers.

© 2018 Great Minds®. eureka-math.org

171

Name _____   Date _____

Count all the objects. Write the number in the first box.

Count the objects that are white. Write that number in the second box.

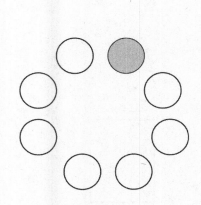

[ ] . One less is [ ] .

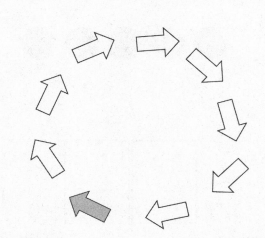

[ ] . One less is [ ] .

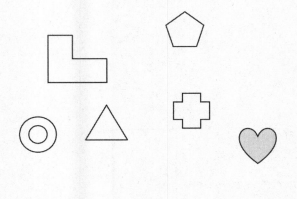

[ ] . One less is [ ] .

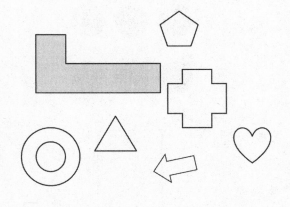

[ ] . One less is [ ] .

Lesson 36:   Arrange, analyze, and draw sequences of quantities that are 1 less configurations other than towers.

© 2018 Great Minds®. eureka-math.org

173

Count and write how many.

Draw 1 less. Write how many.

Count and write how many.

Draw 1 less. Write how many.

Lesson 36:   Arrange, analyze, and draw sequences of quantities that are 1 less configurations other than towers.

EUREKA MATH

Grade K
Module 2

Name _____   Date _____

Sort the shapes.

Shapes with a Curve   Shapes without a Curve

Lesson 1:   Find and describe flat triangles, squares, rectangles, hexagons, and
             circles using informal language without naming.

© 2018 Great Minds®. eureka-math.org

177

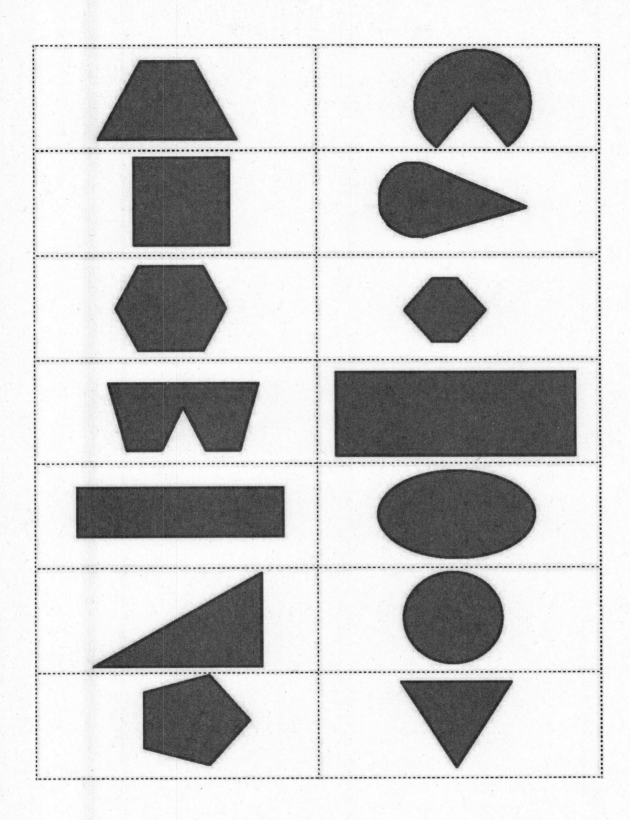

Lesson 1:   Find and describe flat triangles, squares, rectangles, hexagons, and
circles using informal language without naming.

179

Draw a round pizza with your favorite toppings.

With your crayons, show how you would cut the pizza for your family.

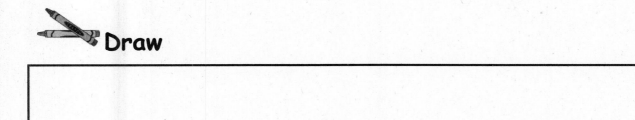 **Draw**

Are your slices the same or different from your partner's slices? Describe the sides and corners
of one of your slices to your partner.

Lesson 2:   Explain decisions about classifications of triangles into categories using
variants and non examples.   Identify shapes as triangles.

© 2018 Great Minds®. eureka-math.org

181

Name _____   Date _____

Find the triangles, and color them blue. Put an X on shapes that are not triangles.

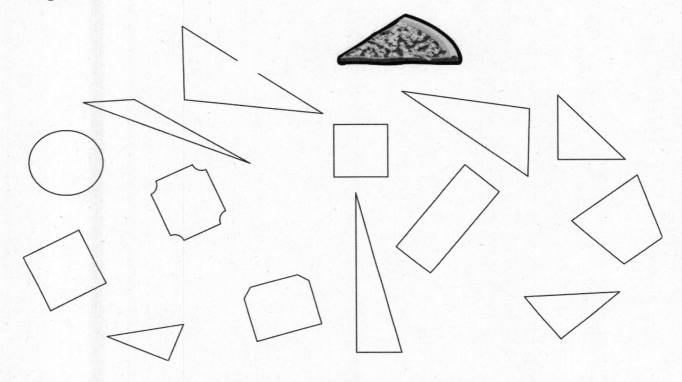

Draw some triangles.

Lesson 2:   Explain decisions about classifications of triangles into categories using variants and non examples. Identify shapes as triangles.

183

Design and draw your own dollar bill.

Draw

How is your dollar the same or different from your partner's? Describe the sides and corners of your dollar to your partner.

Lesson 3:   Explain decisions about classifications of rectangles into categories using variants and non examples. Identify shapes as rectangles.

© 2018 Great Minds®. eureka-math.org

185

Name _____ Date _____

Find the rectangles, and color them red. Put an X on shapes that are not rectangles.

Draw some rectangles.

Lesson 3: Explain decisions about classifications of rectangles into categories
using variants and non examples. Identify shapes as rectangles.

© 2018 Great Minds®. eureka-math.org

187

dot paper

Lesson 3:   Explain decisions about classifications of rectangles into categories
using variants and non examples.   Identify shapes as rectangles.

189

Draw a rocket ship using only triangles and rectangles.

Draw

Trade rocket ships with your partner. Count the triangles. Did you use the same number of triangles in your picture? Count the rectangles. Did you use the same number of rectangles in your picture?

EUREKA
MATH®

Lesson 4:   Explain decisions about classifications of hexagons and circles, and identify them by name. Make observations using variants and non examples

© 2018 Great Minds®. eureka-math.org

191

Name _____   Date _____

Find the circles, and color them green. Find the hexagons, and color them yellow. Put an X on shapes that are not hexagons or circles.

Draw hexagons and circles.

Lesson 4:   Explain decisions about classifications of hexagons and circles, and identify them by name. Make observations using variants and non examples

193

© 2018 Great Minds®. eureka-math.org

Draw yourself.

Draw one thing in the room that is behind you.

Draw

Students stand face-to-face with a partner. Partner A tells Partner B several things he thinks are behind him in the room. Partner B lets Partner A know if he is correct. Students switch roles.

Lesson 5:   Describe and communicate positions of all flat shapes using the words
above, below, beside, in front of, next to, and *behind*.

195

Name _____   Date _____

Cut out all of the shapes, and put them next to your paper with the duck.
Listen to the directions, and glue the objects onto your paper.

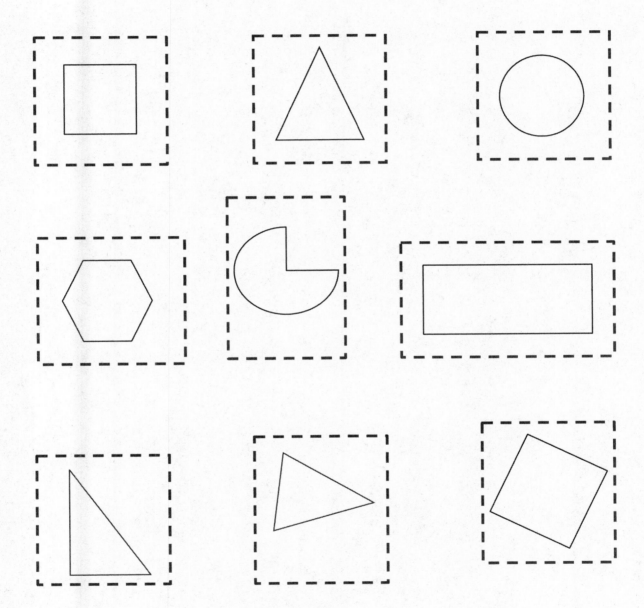

Lesson 5:   Describe and communicate positions of all flat shapes using the words
above, below, beside, in front of, next to, and *behind.*

Name _____   Date _____

Lesson 5:   Describe and communicate positions of all flat shapes using the words
above, below, beside, in front of, next to, and *behind.*

199

EUREKA MATH®

© 2018 Great Minds®. eureka-math.org

Draw a cube.
Draw a ball.
Circle the one that rolls.

 Draw

(Give each pair of students a small ball and a cube). Roll the ball to your partner. Roll the cube to your partner. What happened? Why did the ball and the cube move differently?

Lesson 6:   Find and describe solid shapes using informal language without naming.

© 2018 Great Minds®. eureka-math.org

Name _____   Date _____

Match these objects and solids by drawing a line with your ruler from the object to the solid.

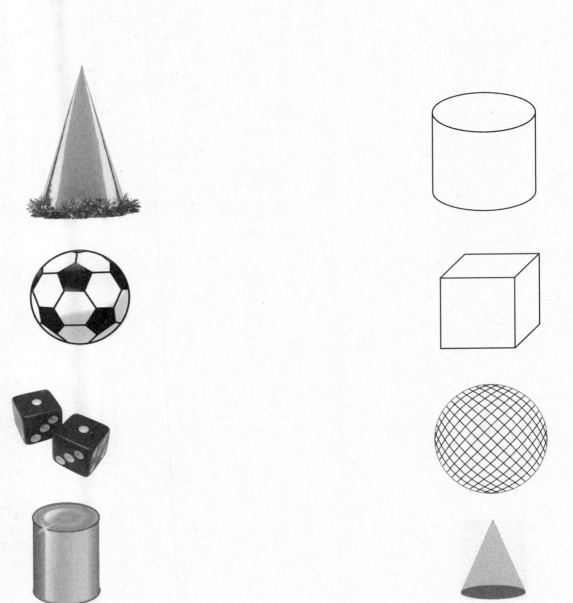

On the back of the paper, draw solid shapes that you see in the classroom.

Lesson 6:   Find and describe solid shapes using informal language without
              naming.

© 2018 Great Minds®. eureka-math.org

203

I am a solid that can roll.

I don't have any corners or edges.

Use your clay to make me.

Then, draw me.

Draw

(Give each student a small piece of modeling clay. Read the riddle. Students make a ball with their clay and then draw the ball.) Look at your solid and your friend's solid. Do your solids look alike?

 Lesson 7:   Explain decisions about classification of solid shapes into categories.   Name the solid shapes.   **205**

© 2018 Great Minds®. eureka-math.org

Name _____ Date _____

Circle the cylinders with red.

Circle the cubes with yellow.

Circle the cones with green.

Circle the spheres with blue.

Lesson 7: Explain decisions about classification of solid shapes into categories. Name the solid shapes.

© 2018 Great Minds®. eureka-math.org

207

EUREKA MATH®

Use your clay to make a cube. Make a cylinder. Make a sphere.

Make a cone.

Draw a sphere above a cone.

 Draw

(Give each student a small piece of modeling clay. Read the directions.)

Lesson 8:   Describe and communicate positions of all solid shapes using the
words *above, below, beside, in front of, next to,* and *behind.*

© 2018 Great Minds®. eureka-math.org

209

Name _____   Date _____

Lesson 8:   Describe and communicate positions of all solid shapes using the
words *above, below, beside, in front of, next to,* and *behind.*

© 2018 Great Minds®. eureka-math.org

EUREKA
MATH

211

Directions: Read to students.

Paste the sphere **above** the train.

Paste the cube **behind** the train.

Paste the cylinder **in front of** the train.

Paste the cone **below** the train.

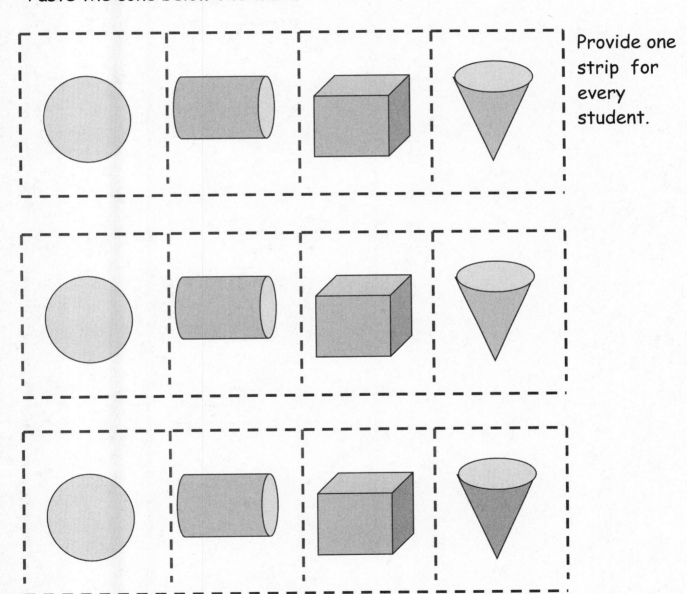

Provide one strip for every student.

Lesson 8:   Describe and communicate positions of all solid shapes using the words *above, below, beside, in front of, next to,* and *behind.*

© 2018 Great Minds®. eureka-math.org

213

Draw a shape.
Make a solid with your clay that has the shape you drew as one of its faces.

 Draw

(Give each student a small piece of modeling clay.) Share your work with a partner.

EUREKA MATH®

Lesson 9:   Identify and sort shapes as two-dimensional or three-dimensional, and recognize two-dimensional and three-dimensional shapes in different orientations and sizes.

© 2018 Great Minds®. eureka-math.org

215

Name _____   Date _____

Circle the pictures of the flat shapes with red. Circle the pictures of the solid shapes with green.

Lesson 9:   Identify and sort shapes as two-dimensional or three-dimensional, and recognize two-dimensional and three-dimensional shapes in different orientations and sizes.

© 2018 Great Minds®. eureka-math.org

217

Name _____   Date _____

These are (____).   These are not (____).

_____

work mat

Lesson 10:   Culminating task—collaborative groups create displays of different
               flat shapes with examples, non-examples, and a corresponding solid
               shape.

219

Credits

Great Minds® has made every effort to obtain permission for the reprinting of all copyrighted material. If any owner of copyrighted material is not acknowledged herein, please contact Great Minds for proper acknowledgment in all future editions and reprints of this module.

- Lesson 1, p. 5: Rabbit photo credit: Eric Isselee / Shutterstock.com